Sussex Privies

by

David Arscott

COUNTRYSIDE BOOKS

NEWBURY · BERKSHIRE

First published 1998
© David Arscott 1998
Reprinted 2000

COUNTRYSIDE BOOKS
3 Catherine Road
Newbury, Berkshire

To view our complete range of books,
please visit us at
www.countrysidebooks.co.uk

ISBN 1 85306 534 X

Produced through MRM Associates Ltd., Reading
Typeset by Techniset Typesetters, Merseyside
Printed by Woolnough Bookbinding Ltd., Irthlingborough

CONTENTS

FOREWORD

It is everyone's guilty secret, our shameful personal history of subterranean gurgles, searing explosions and wafting odours; of locked doors and furtive wipings; of urgent disburdenings delayed too long. To bear a surname such as mine is constantly to be aware of a shared unease about these bodily functions and the parts associated with them: overt schoolboy jests about that first syllable are far less common than an embarrassed refusal to pronounce it properly. Excretion may be our common lot, but we would each like to be thought an exception to the rule.

This shy distaste has doubtless played a part in the sorry vandalism which has destroyed so many of our garden privies. Although the less rickety have often survived as tool sheds and coal bunkers, our current rage for conservation came too late to save much of the furniture inside. Following long and winding rutted tracks through remote farm gates, I have certainly found some endearing little buildings which only the elements have thought to slight, but those wooden planks with the holes cut in them have usually long since fuelled someone's bonfire.

If it was unreasonable to expect many primitive throw-backs in the towns, with their ubiquitous flush lavatories and mains drainage, I nevertheless suffered one of my greatest disappointments at Newhaven, where, I was told, the cell door of the former police house had been fixed to the outside loo. The thought of finding a spy-hole on a building whose very name speaks privacy did rather tickle me, but that, too, had gone.

These frustrations might have proved terminally damaging to the project had it not been for the amazing public response to my request for tales of 'the bad old days'. That long trip up the garden path, all too often in the cold and dark, is not easily forgotten, and the recovered memories were, as you will discover, colourful and varied. 'I hope you're not going to make fun of privies,' said one woman anxiously. 'They were a godsend at the time. Just think of the alternative.'

4

A common fate. This sturdy sand-stone privy at Hardham is now a garden store . . .

. . . like this one. Surely that large window wasn't original?

Well, of course there are plenty of laughs in these pages – together with (the more shockable be warned) a few inevitable impolite expressions. To the young it is scarcely credible that many a Sussex man and woman 'performed' in a draughty outhouse less than half a lifetime ago. Several minutes into an interview on local radio, I realised with a start that the twenty-something presenter had no idea what I was talking about – that older generations had to go in a bucket, using squares of news-paper that hung on a nail in the wall. What, no chain? He obviously found it highly amusing.

But there is, although it needs no solemn overstating, a serious side to privy-hunting, too. The disposal of human waste forms an essential part of social history, and in these pages we are given a rare and revealing glimpse of the underside of our Sussex story.

DAVID ARSCOTT

The author outside his own privy at Westmeston – now, like so many, nothing but a tool shed.

[1]

THE PASSING YEARS

After the Roman left old Rye
And out from Severn strode,
The needful Sussex peasant
Used the fields to shed his load.

Just as the roads they built were later to become overgrown and, in many cases, completely forgotten, so the standards of hygiene introduced to Sussex by the Romans were soon nothing but a memory.

While the vast majority of ancient Britons never shared the luxuries of their imperial masters, many would have come close enough to gawp at the bath suites with their fancy mosaics. Fishbourne, that great palace on the edge of Chichester Harbour, must certainly have had comfortable lavatories, flushed by water from the cold plunge into an efficient sewer system; unfortunately, less than half the site remains today, and those facilities, draining into the harbour, would have been on lower ground now inaccessible to archaeologists.

At Bignor, tiling has been found in a field close to the bath suite with its wonderful Medusa mosaic, and this – again, at the lowest level – was almost certainly the site of the villa's lavatories. There are, alas, no plans to excavate the area, so we can only imagine the wealthy villa owner perched comfortably on his marble seat above a trough of flowing water, perhaps overlooked by those household deities Stercutius (the god of manure) and Cloacina (the goddess of sewers).

The business over, he would have cleaned himself by means of a sponge on a stick, a useful device which was, however, open to misinterpretation by those not yet accustomed to the ways of higher civilisation: the philosopher Seneca tells us of a German

7

The garderobe, or latrine, shaft at Michelham Priory discharged into the moat. Fortunately the water was free-flowing.

who mistook this implement for a tempting skewered foodstuff, with fatal consequences.

The Roman period was followed by what we might reasonably call the Dank Ages. We hear nothing more about such matters until medieval times, and then most of what we know will make the sensitive shudder.

'Beware of draughty privys and of pyssynne in draughts,' warns Andrew Boorde, Henry VIII's doctor and a man we can claim for Sussex (he was born near Cuckfield and later lived at the Old Mint House by Pevensey Castle), 'and permyt no common pyssing place about the house.'

Boorde, a jovial scholar, made a connection between health and cleanliness which seems obvious today, but which no doubt surprised his contemporaries.

'Let the common house of easement to be over some water or else elongated from the house,' he went on. 'Beware of emptying pysse pottes, and pyssing in chymnes.'

Relieving yourself in a grate conveniently full of wood ashes must have been a real temptation when caught short. As for the apparently commonplace habit of emptying chamber pots from upstairs windows, it was supposed to be accompanied by the cry of 'Gardy loo!' (from the French *gardez l'eau*) in the hope that passers-by might take avoiding action.

Henry VIII himself used a 'close stool', a box with a padded seat over a pewter pot which his Groom of the Stool had to empty, but only the wealthy could expect such a service. If the inhabitants of castles and monasteries had more comforts than the common man and woman, their 'garderobes' (or wardrobes, quaint euphemism) would surely make 20th-century sybarites cringe, although John Guy, in *Castles in Sussex*, offers a corrective to excessive squeamishness: 'Castle interiors were very much more comfortable than we might imagine,' he writes, 'and included such luxuries as flush toilets, running water, brightly-painted walls and, in some extreme cases, double glazing.'

The monasteries, of course, were largely destroyed in Henry's

The latrine shaft at Michelham Priory from the inside, now covered by a
safety grille.

Going, going, gone . . . This attractive stretch of water at Michelham Priory is the course taken by the monks' waste on its way to the nearby River Cuckmere.

time, and in Sussex we have only the outlines of 'reredorters' – dormitory annexes which housed the lavatories. At both Bayham Abbey and Michelham Priory, however, we can trace the ditch which carried the waste away to nearby rivers, while the thick walls of the late 14th-century gate tower at Michelham incorporate latrines which hang over the moat. This was, and is (should you be thinking of some very unpleasant long-term stagnation), constantly moving water.

Garderobes were usually approached by a right-angled passage which served to trap some of the offensive smells, and they were sometimes built close to chimneys for warmth. At Amberley Castle we have what Guy regards as 'two of the finest latrines to have survived from the Middle Ages', with a possibly unique arrangement of shafts on either side of the garderobe tower. Bodiam, that fairy tale creation, has no fewer than 28 separate latrines, each with its own drain shaft to the moat, and there are other examples at Lewes and Pevensey.

And if there was no convenient moat or ditch to swirl the stuff away? Indoors, ordinary folk would have used the ubiquitous 'piss pot', but wealthier people had other arrangements. The Weald and Downland Open Air Museum at Singleton has an early 15th-century Wealden hall-house ('Bayleaf') complete with a garderobe which juts out over the garden. The guidebook, admitting that its reconstruction is partly conjectural, adds: 'It is not clear whether it originally emptied straight into an open cess-pit or whether there may have been some kind of covered conduit, but the contents were used as garden manure.' Doubtless the well-to-do owner had someone else carry out that particular duty.

The first valve water closet was devised in 1596 by Queen Elizabeth's godson, Sir John Harington, but for some reason it never caught on.

'In the privy that annoys you,' he wrote, in a fine piece of advertising copy, 'first cause a cistern of lead containing a barrel or upward to be placed either behind the seat or in any

Amberley Castle is unusual in having a pair of latrine shafts, their arches springing from either side of the garderobe tower.

Where the well-to-do went in medieval times. This garderobe, in a corner of the upstairs apartment at the Bayleaf hall-house (now reconstructed at the Weald and Downland Open Air Museum, Singleton), juts out over the garden.

And an exterior view. The museum guidebook says there may have been a covered conduit over the pit, and the contents were eventually used as garden manure.

place either in the room or above it, from whence water may, by a small pipe of lead of an inch be conveyed under the seat in the hinder part thereof (but quite out of sight); to which pipe you must have a cock or a washer to yield water with some pretty strength when you would let it in.'

The device was reinvented by the watchmaker Alexander Cummings in 1775, and this time proved more successful, but it wasn't until the Victorian period that real improvements were made to general standards of hygiene. The Public Health Act of 1848, for instance, ruled that a fixed sanitary arrangement (albeit that it could be something as rudimentary as a bucket) must be installed in every household, and a second such act during Disraeli's premiership in 1875 created a nationwide system of sanitary authorities responsible for water supply and sewerage.

At the International Exhibition of Hygiene at South Kensington in 1884 there was a competition for the best water closet, although only three of the thirty entries managed to pass the test set for them: to remove in a single flushing a deposit of ten small potatoes, some sponge and four thin sheets of paper. A bevy of sanitary engineers had by now set themselves the task of cleaning up the act, and in so doing left their names to posterity: Doulton, Twyford, Shanks, Crapper.

For some poor souls, however, the age of universal flush lavatories came to Sussex too late.

[2]

VICTORIAN VILENESS

'A disgrace to any civilised people'

In June 1849, the very month that the civil engineer Edward Cressy arrived in the town to begin an official, and gruesomely revealing, sanitary inspection, the first victims of Brighton's hidden cholera outbreak fell ill and died.

Cressy's eventual report makes it clear that this was an epidemic waiting to happen, the finger being pointed at the disgusting state of the town's privies and the cesspools into which they drained. The population had grown from 7,000 to 60,000 in only fifty years (it was to top 100,000 by the end of the century), and in the poorer districts the houses were clustered together with little regard for even the most basic hygiene.

'At no. 118,' he found, reporting on Woburn Place, 'is a double privy and a cow lair which drains into an old well, which is never emptied. This double privy is used by most of the dwellers in this street, from their own being in so bad a state.'

Worse, these privies were sited *above* the houses, so that liquid excrement often flooded into them from overflowing or imperfect cesspools.

As Roy Grant has revealed in his paper *Observations on Brighton's Cholera Outbreak of 1849*, the authorities hushed up the epidemic for fear of its effect on the town's tourist trade. Although the exact number of deaths will never be known, however, small crosses next to names of the dead in the St Nicholas parish register offer a coded reference to the disease, indicating that there were all of 175 victims in that first year. By 1854, the cross is replaced by a large C.

Cressy's report wasn't the first to condemn the vileness of Brighton, which in 1849 had something like forty miles of streets but only six miles of sewers. Dr G. S. Jenks, who contributed to Edwin Chadwick's seminal 1842 *Report on the Sanitary Condition of the Labouring Population of Great Britain*, found 'a multitude of ill-contrived narrow, thronged and pent-up lanes, courts and alleys' peopled by 'wretched and sickly-looking inmates'.

During heavy rain the privies – 'mostly in dire need of empty-ing' – would overflow and corrupt the well water. In the lower areas the houses often had 'material' saturating the foundations of the houses and oozing through the walls.

'Owing to the imperfect and insufficient drainage of the town,' read Chadwick's report, 'the inhabitants are compelled to have recourse to numerous cesspools as receptacles of super-abundant water and refuse of all kinds, and to save the inconven-ience of frequently emptying them, they dig below the hard coombe rock till they come to the shingles, where the liquid filth drains away.

'The consequence is inevitable: the springs in the lower part of the town are contaminated.'

Dr William Kebbell, who in 1848 described the conditions in the poorer parts of Brighton as 'a disgrace to any civilised people', estimated that more than 800 deaths a year were caused by diseases entirely preventable by proper sanitary measures.

Despite all this irrefutable evidence, local officials who feared losing their clout to a centralised health authority managed to block any serious improvement of Brighton's drainage system until the 1870s, when Sir John Hawkshaw's great intercepting sewer was built – a main trunk route which took the outpourings of all the smaller sewers and carried the wastewater out to sea. A number of pumping stations were constructed (including what is now the British Engineerium off Nevill Road, Hove) to provide fresh drinking water and, just as important, to flush the sewers.

We shouldn't assume from this general squalor that most poor families made no attempt to keep themselves and their houses decent. Many of the contributors to this book, though they

certainly never knew the horrors of Victorian urban depriva-
tion, have stressed the strenuous attempts by their mothers to
keep the privy spotless (the seat 'scrubbed as clean as our bread
board,' one recalled memorably), and we have no reason to
believe that things were different for earlier generations.
Conditions simply overwhelmed them.

Nor was the great metropolis of Brighton the only Sussex town
to be ravaged by germs that thrived on overcrowding and poor
drainage. In September 1880 typhoid (or enteric) fever broke
out in Uckfield. It wasn't the first visitation of the disease locally,
but it was certainly the worst: thirty-eight people fell ill and five
of them died.

W. H. Power, commissioned to investigate the causes on
behalf of the Local Government Board, wrote a devastating
report: the town had had sewers for only twenty-five years, in
which time few had ever been inspected and cleaned; these
sewers were not ventilated ('the Uckfield sewers ventilate very
generally into the dwellings and closets of the place'); most
houses drained directly into them, 'a trap on the house drain
shutting it off from the sewer being almost unknown in
Uckfield'; and the wells were contaminated.

'The water supply', Power wrote, 'is from wells commonly
situated close to, and even under, dwellings which in their turn
occupy sites which have for generations been fouled by soakage
from privies, cesspools, ashpits, defective drains and the like.

'At ordinary times, no doubt, the wells are polluted to a
greater or lesser extent by nothing worse than non-specific
organic refuse, which has soaked into them from the shallow
filth-saturated soil of the town, or has reached them more
directly, perhaps through fissures in the rock, from leaky drains,
old cesspools or privy pits.'

When typhoid was rife 'the excrement of enteric fever
patients' was added to this revolting stew, with predictable
results.

Where the drains emptied into a pit, local brickmakers were

employed to cart off the waste, but they carried out the work only when it suited them, and then only once or twice a year. As for the outside privies, most were over a drain, but very few had any water supply. Their 'inadequate and irregular' flushing depended on an occasional pail of water cast into them by the tenants, who chiefly used refuse water for the purpose 'since persons will not be at the trouble of draining well water from a depth of 50 feet merely for the sake of casting it at once down a closet.'

There's an echo of this in a memory of Lewes in the 1930s ('The gardens were separated by little fences, and each morning you'd see the mums going to the privy to empty the slop buckets'), although the words of another correspondent, but referring to the same town during the same era, ('A visit to the loo meant a bucket of water and the Jeyes fluid') remind us of the kind of thing our forebears *didn't* have: a solution of chloride of lime or quicklime was the contemporary state of the art.

In his history of Uckfield Parochial School, *Payment by Results*, Simon Wright shows that these lessons were not quickly learned. Diphtheria was another local scourge, and several children succumbed to it during the 1880s and early 1890s. The Medical Officer of Health, Dr E. F. Fussell ('it is scarcely too much to describe him as a Sherlock Holmes of the world of hygiene'), was a tireless campaigner for better sanitation, highlighting deficiencies in the sewer system and urging the headmaster to consult a map of the drains: 'It, or a copy, should hang in your house, I think.'

W. H. Power had an interesting comparison to make between the sanitation of poor and better class houses. The latter, it was true, often had a water supply from cisterns fed by rain spouts, or filled by pumping, but they had inside lavatories directly above the open sewers, and 'their drain connections are numerous and complex'. Both the smell and the health risk must have been uncomfortably high.

The poorer people, he claimed, were luckier in this respect, since they 'have not always indoor sinks, and their pan closets are commonly separate from the dwelling.'

Here we find our very first salute to the down-the-garden-path kind of privy celebrated in this book. We may smile at its ramshackleness ('Ours blew over once or twice' is a typical memory, 'and the wind often rocked it about while we were sitting inside') and grimace at the *in*conveniences involved in using it. We may allow that its hygiene rating is often far from ideal.

It lies, on the other hand, safely away from the house in the clean, fresh and invigorating air – a little retreat, moreover, for escaping the hurly-burly of life and glorying in the close proximity of bird, beast and flower.

Spending a penny black ... This privy at the junction of Western Road and Cuckfield Road in Hurstpierpoint must have been a mite inhibiting to use when someone was posting a letter. It was so prominent that it gave the spot the (unofficial) name of Shithouse Corner. The privy remains, though now hidden behind a fence, but the letterbox – the only VR type in the village – has been moved across the road.

[3]

RATS AND OTHER COMPANIONS

'It was quiet down there on a summer morning, sitting
there with the door open watching the honey-bees buzz
around from blossom to blossom in the flower bed and
listening to the old blackbird filling the air with song
from the top of the holly tree.'

Jim Copper's fond memory, as retold by his son Bob, finds an echo in many a Sussex heart. Time and again I have heard older people enthuse about the pleasure of defecating *al fresco*, the sun beating full on one's face through the open door, a soft breeze bringing in the sweet fragrance of may, syringa and rose.

And at night-time, too. Here's Michael Ridley of Peacehaven: 'The ability to sit with the door open, looking at the stars and the distant lights of the coast, was an absolute joy.'

An enthusiast for Harry Cawkell's 'Nature's Way' column in the *Evening Argus*, I was delighted, while researching this book, to find him devoting one of his weekly spots to the wildlife spied from his privy. The opening paragraphs give the overall flavour of the thing:

I have a pygmy in my loo – and I bet there aren't many people who can match that.

The loo, I must explain, is one of the old-fashioned outdoor type, not exactly at the bottom of the garden but far enough away from the house to be a rather inconvenient convenience when the weather is wet and windy.

It's pleasant, however, on a fine summer's day. The situation being quite rural, one can then leave the door ajar and enjoy the living world, communing with nature rather than a newspaper.

And because one is in shadow it's like being in a hide, and the

A preserved privy at Grovelands farm, Hailsham, which should see better days yet. The town council now owns the farm and is determined to retain its character. Betty Driver remembers sharing this privy with rats.

birds and mammals behave as if one wasn't there.

The result is that over the years I've had a number of visitors. Mostly they've been birds. On one occasion a half-grown grey partridge strolled in, peered around, decided there was nothing edible on offer, and strolled out again.

Swallows have often carried out a reconnaissance, perhaps checking whether the loo would make a desirable nesting site.

The annoying thing is when they then try to exit via a small, non-opening window on one side and a lot of arm-waving is necessary to steer them back out of the door.

Starlings have been frequent intruders, charging about knocking down the spider webs, and house sparrows mysteriously seem to be able to get in – but not out – even when the door is closed. Invariably, when freed, they leave behind their visiting cards.

Wrens and warblers (willow warbler and blackcap) have always been welcome, but not a blackbird one year that seemed to regard the loo as a sanctuary when being pursued by the owner of the territory.

Gulls, magpies, crows and red-legged partridges are among the heavies that regard the roof as a convenient perching place, while in contrast one bitter winter evening I found a tiny, half-frozen goldcrest crouched against the door, seeking what shelter it could.

Fortunately it revived in the warmth of my kitchen, where I roosted it overnight.

Mammal visitors have included a 'mixy' rabbit that blundered in one afternoon; a young fox that peered round the door frame and then fled; and frequent woodmice, plus a suspected yellow-necked mouse.

That tiny interloper of the first paragraph, in case you're still wondering, was a 5cm-long pygmy shrew.

But for everyone who gloried in the abundant wildlife of the privy, there were a dozen shivering wrecks appalled at the thought of encounters with scurrying and scuttling things.

'Oh, those huge garden spiders!' remembered Pansy James. 'Probably why I still dread each August approaching, when they tend to come inside. In the summer time there was another unmentionable hazard.'

I realised, when putting this material together, that I never *had* asked her to explain that last remark, but I'm happy to let it stand as a reminder that we each of us have our own particular 'hazard'. For Betty Driver of Stanmer it was frogs:

'Hated the things! When you went out there in the spring they were all over the path, and you had to step carefully round them so that you didn't squash them under your feet.'

Eileen Parsons, now of Shoreham but then a child in Henfield, failed to take avoiding action on one occasion, the injured party being rather larger than a frog.

'I remember one particularly dark evening, visiting the "accommodation" (what was wrong with the lantern or torch on this occasion I don't know), and to my horror on entering I stepped on something round and squashy.

'Yes, you've guessed: a hedgehog! I'm not sure who was more scared, myself or the hedgehog. What a good thing it wasn't on the seat.'

Darkness was a problem for Norma Noakes one evening at her nan's privy in Brighton.

'I was just sitting down when I felt something brushing against my legs. I was only seven or eight at the time. I rushed into the house, terrified, leaving the door wide open – and in behind me came a ferret! I think it had escaped from a garden close by.'

Norah Browning lived in Hurstpierpoint High Street, with an outdoors back-to-back loo that enjoyed the luxury (as with many town privies) of being on the main drainage system. That didn't deter the wildlife, of course, and 'my father fitted the box underneath with broken glass to keep the rats away'.

Public Enemy Number One, surely. And yet, although I

When Jim Driver and his wife Betty moved into one of the Stanmer Park gate-houses in 1956, their lavatory was this privy up the garden path: they used it until 1963. The pretty terracotta finial on the gable end is a gracenote added by Betty.

Something to keep the Devil (and perhaps the wildlife?) at bay.

found rat-haters in plenty throughout Sussex, I was to come across a remarkable case of live-and-let-live.

Betty Coleman's home was a farm at Hailsham (see page 23) 'and I used to hang bread and fat for the birds on the big lilac bush outside the privy,' she told me. 'The rats liked to feed on that. You'd see them sitting in the bush, tucking in.

'There was a square hole in the privy door, and at night they'd get inside and run around the eaves above your head.

'A rat will attack you if it's cornered, but I was never troubled in there. They had the run of the place. I brought up my step-children on the farm, and they never complained, either.

'I used to sit in there and sing away!'

That fine actor Dirk Bogarde was brought up at Lullington, near Alfriston, an idyllic childhood vividly portrayed in the first part of his autobiography, *A Postillion Struck by Lightning*. He knew what it was to share the facilities with wildlife:

Round the cottage was a rickety wooden fence with bits of wire and an old bedstead stuck in it, and some apple trees and the privy with its roof of ivy and honeysuckle and a big elderberry. The privy had no door, so you just sat there and looked into the ivy; no one could see you through it, but you could see them coming along the little path and so you were able to shout out and tell them not to in time. It was really quite useful. And better than a door really, because that made it rather dark and a bit nasty inside. And once a bat got in there after Lally had closed it and she screamed and screamed and had a 'turn'. So we left off the door for summer and just sort of propped it up in the winter, to stop the snow drifting and making the seats wet.

There were three seats, like the Bears'. A little low one, a medium one, and the grown-up one. The wood was white and shining where we used to scrub it, and the knots were all hard and sticking up. No one ever used the smallest one. We had the paper and old comics and catalogues for reading in that; and the medium one just had a new tin bucket in it with matches and candles for the candlestick which stood on a bracket by the paper roll, and a cardboard tin of pink carbolic.

There were lids to all the seats, with wooden handles, and they had to be scrubbed too – but not as often as the seats; which was every day and a bit boring. Sometimes at night it was rather nice to go there down the path in the dark, with the candle guttering in the candlestick, and shadows leaping and fluttering all around and the ivy glossy where the golden light caught it. Sometimes little beady eyes gleamed in at you and vanished; and you could hear scurrying sounds and the tiny squeaks of voles and mice; and once a hare hopped straight into the doorway and sat up and looked at me for quite a long time, which was fearfully embarrassing, until I threw the carbolic tin at him and he hopped off again.

Domestic creatures could be a self-inflicted punishment, as Michael Ridley recalled: 'One of the major problems was a chicken – actually a fiery little rooster called Roger, who was allowed to roam freely round the garden.

'This was a particular problem when trying to get to the toilet, or when trying to get back to the house afterwards. To solve this a number of sticks were stationed around the garden to be grabbed and used to keep Roger at bay. A tennis racquet was kept in the toilet itself as added protection.'

For Dot Collyer of Bognor Regis the fear, during her childhood on a farm, was of one of the family's geese.

'He took a dislike to me for some reason. If I needed to pay a visit I was all right during the day, because I could see it, but when it was dark, that was another matter. I either took somebody with me, or I would have a few accidents. This particular goose would go for my legs. My parents couldn't understand my fear.

'But it didn't end there. My grandparents also had an outside loo, but theirs was a very long way from their house, because they had a big garden. It was surrounded by tall fir trees, so very spooky at night. There was no electricity, of course. No

This attractive little privy at Gibraltar Farm, Firle, has been left unspoiled by the Barnes family. A step inside reveals a quality ceramic bowl with a proud Dombey trademark.

The longest journey ... The author fondly recalls his discovery of this little sandstone privy because his search took him off the metalled roads and along miles of winding, bumpy tracks in the western Weald until – a blessed miracle! – he came over a rise and saw below him the most beautiful tile-hung medieval farmhouse.

one went down on their own at night.'

Betty Greening knew that feeling well.

'I was born and brought up at the White Hart in Stopham. The house was on one side of the road and the garden on the other, so you had to cross the road to reach the closet. It was very spooky if you needed to go after dark. When I was small I wouldn't go on my own, even by the light of our candle lantern.'

Animals, then, weren't the only concern.

'When we stayed with our grandparents,' said Olive Morgan, 'theirs was open to all the elements. The wooden panels had parted over the course of time. Needless to say, you waited until you were desperate before you went there.'

'Until the very last minute,' agreed Betty Driver. 'And then you were so close to bursting that you didn't notice how dark or wet or cold it was.'

'In the winter,' said Jean Lawrence, 'we took either torch, candle or a night-light to see what we were doing, but worst of all was the fact that when you were sitting down the wind blew up into your bottom.'

'Ours at Beddingham had a large square hole in the back, perhaps for ventilation,' Ron White told me. 'When it snowed you knew all about it! Eventually we plugged it with a Lipton's Tea box.'

'Our outside loo in Seaford faced north,' remembered James Payne, 'so a visit in the depths of winter wasn't appreciated. The cistern water usually froze, and my father fixed a shelf on which was placed a night-light, which was supposed to keep it running. If it didn't you were back to a bucket of water.'

'You left by the back door,' was Elizabeth Skinner's memory, 'through the roofed-in well-house, round the corner of the cottage – where the candle invariably blew out – and through the coal shed to the sanctuary. It was bliss to be given an electric torch for my seventh birthday.'

'It could be scary up there at night,' someone who identified herself only as Nibs told me on a local radio phone-in. She'd lived in the wilds of Herstmonceux. 'We were told that if you banged on the door three times the fairies wouldn't get you.'

'Dark evenings were the worst time,' wrote Norah Browning, 'when, armed with a candle, hand cupped round the flame, one would venture to the loo, which seemed far away from civilisation. I always sat with the door open in case a bogeyman came.'

The answer was to have company, and Myrtle Hancock told me of a very long-suffering gentleman.

'My father was always very good when my next sister in the family needed to go to the loo in winter time, when it was dark. It seemed she was more usually needing this than anyone else, and my dad would sigh and put his boots back on, light a lantern (this would be about 1920) and conduct a jaunt up the garden, where he would patiently wait under the yew trees.

'He also performed this service in later years for my daughter, who was like my sister – most likely to be wishing for a visit up there when it was dark and cold, or raining.

'I may add that we didn't have up-to-date indoor facilities until 1965.'

Pauline Lavington played the guardian angel role when she was a child in Nutley, but she was (and understandably) somewhat less reliable than Mrs Hancock's father.

'When I was five and my older sister was twelve she was frightened of the dark, so I had to go up the garden path with her, and all the time she was in there I had to stand outside singing.

'I wasn't allowed in with her. Sometimes I'd get fed up and stop singing, and then she'd come running out in a panic with her knickers round her ankles.'

Oh, the agonies: not only palpitating fear but an all too easily aroused sense of toe-curling shame.

[4]

THIN WALLS, RED CHEEKS

'We had a three-holer at Upper Parrock Farm,
Coleman's Hatch, and when we needed
to take that trip up the garden path we'd
call out "Anyone else like to come?"'

Lack of privacy was one of the worst aspects of the garden privy. Michael Balcon, who furnished the above memory, was at least in the company of friends and family. All too often that trek to the loo was undertaken under what seemed to be the prying gaze of prurient neighbours.

Evelyn Dedman of De Montfort Road in Lewes had good reason to be offended.

'When we moved into this house in 1961 we found the outside toilet had originally been a two-holer. It had been adapted to a flush toilet, but it still had the original wooden seat with two holes, one larger than the other.

'My elder daughter was then just ten months old, so when I went out there I had to take her with me, so she would sit on the little hole. We had a nosey old neighbour who would stand outside his door, and whenever I had visited the loo he would say: "You've been there a long time, Eve."

'As a young wife and mother, I was so embarrassed I'd hold on as long as possible before a visit.'

In many cases, of course, neighbours must have been equally uneasy about the situation.

'Unfortunately,' recalled Norah Browning of her Hurstpierpoint experience, 'the door was exactly opposite the door of the next house, so on occasions we both emerged at the same time.'

Jean Lawrence was born in Sun Street, Lewes, in 1930:

'The privy was at the bottom of the garden, backing onto the

The good companion. The author's 2-year-old daughter Beth, seen here outside a one-holer at Hurstpierpoint, accompanied him all over the county on his privy hunt.

neighbours'. I remember the wooden seat was wall-to-wall, with just one hole in the middle. We used to have to take buckets of water to throw down after use.

'If one of the neighbours was in theirs a conversation took place between them and us. The walls were thin, you see. Makes me cringe now, when I think about it.'

Olive Morgan lived in a small flint cottage in the heart of the county:

'The particular place was a combination of three for the three cottages, so close neighbourliness was essential. The walls between didn't reach the sloping roof at the highest point.

'One day when my father was in ours he heard a noise coming from the next one. He promptly said: "That was a good one, Emma!"

'Imagine his surprise when he came out and found a stranger, who was staying with a neighbour, emerge from the other door.'

Jim Copper's way of dealing with possible intrusions will be familiar to many:

'We lived in the end cottage of a row of four, and the privies, or dunnicks as we called them, were halfway down the garden path about twenty yards from the back doors.

'That was quite a jaunt, particularly on a rainy day, and if while you were there you heard the click of the heavy old lift-latch and the familiar creak of the hinges as someone came through the back door, setting out on the same job as you were on yourself, you would suddenly burst out singing at the top of your voice to let them know you were there, and to save them an unnecessary journey.

'It didn't matter what the song was about so long as you made yourself heard. It was an act of common courtesy.'

Although I never came across a case of sexual consummation during my researches, Eileen (as befits the medium, she didn't fuss with her surname) told me the following story on a radio phone-in.

'We had a shared loo with next door, and there was a small hole in the wall between.

'I was in there one time when the lad next door came in and tried to make a date with me. Just like that! While I was on the loo!

'I said No.'

John Goddard, now of Brighton, was one of my most energetic contributors, comparing and contrasting a host of privies he had experienced over the years. To hear his story of being admitted to hospital, a young lad who had never known modern sanitary arrangements, is to remember how recently the Sussex poor were made to feel their poverty.

'I must have been around nine or ten,' he wrote, 'when I became ill and was admitted to the Princess Alice Hospital at Eastbourne.

'There were three of us boys, and we were called The Three Musketeers. We all had appendicitis.

'I hadn't used a water toilet, not used electric light nor had a bath in a big bath (we had a tin bath and hot water on the stove). It was an elaborate affair, with a big toilet seat, wooden, and a large cistern and a big chain, which for me didn't seem to work at all in the first place. I hadn't seen one before and no one told me what to do with it. Very embarrassing.

'I must admit there was a friendly Irish nurse who was a great help. It was the first time I'd ever tasted Irish stew, and I had a double helping. Somehow I had to pay for that, and I dreaded going to that loo as I still didn't know how to use it.'

As the world grew more sophisticated, those still using privies found the very fact of their relative primitiveness an embarrassment.

'We lived at Falmer,' said Ruth Whitehead, 'and we had privies right up until the 1960s. The Brighton Council men came round to empty the buckets on Mondays and Fridays, and it made a disgusting smell. The bucket men would come in for a cup of tea.

'What made it harder was the fact that the university had been built next door with all mod cons.'

Ann, a local radio caller, had become accustomed to something better when she found herself confronted by something basic – though not, one imagines, the original garderobe – at Lewes Castle.

'This was about thirty years ago, and it was three weeks before my daughter was born. You know how desperate pregnant mothers can get. We asked an attendant where the lavatories were, and he unlocked two cranking doors and inside there were three holes in a row on a wooden plinth.

'I did use it, but I needed to go again half an hour later, and I just couldn't bring myself to go through all that again!'

If you had used a privy at home for years without number you had at least grown acclimatised to the attendant embarrassments. For visitors it could be much more difficult – and that, in turn, only served to make your own cheeks redder.

Olive Morgan recalled her father's exasperation with sophisticates from the capital – and an 'invention' which, if it had survived intact, would surely have had future archaeologists thinking that they had turned up a previously unknown type of privy.

'We often had people from London to stay. They hated these loos so much that, rather than use them, they would walk to the Eight Bells pub or even get on the bus and go to Eastbourne.

'After a few times of this, my father managed to persuade the estate manager to get us an Elsan so that he could build a wooden cubicle away from the others.

'This done, he found a large spring which he fastened to the roof. He then tied a thin rope to it and hung it down inside so that they could tug on it and believe they had actually flushed the loo!'

All alone am I . . . From the recreation ground at Steyning you can enjoy the fine sight of a long row of back-to-back privies in the rear gardens of Charlton Street. This is a conservation area, which means that all buildings should be protected, but this strangely shaped privy has at some time been cruelly deprived of its siamese twin.

[5]

FROM PIT TO BUCKET

'There's a lot of fine point to puttin' up a first class
privy that the average man don't think about. It's
no job for an amachoor, take my word on it.'
Charles Sale, *The Specialist*

My copy of the quirky little book quoted above carries the information that it was first printed in England in 1930, and that forty years later it was in its forty-fourth impression, having sold 648,000 copies. Assuming its popularity to have continued undiminished, it must by now have clocked up the magic million. I feel rather ashamed to say that, unlike most of the people I was to meet during my investigations, I had never come across it before. I am now happy to recommend it, not only for its humour but because it gives a very good idea of the structure of the common-or-garden privy.

Here's the narrator, 'champion privy builder' Lem Putt, on door openings:

'Now' I sez, 'how do you want that door to swing? Openin' in or out?' He said he didn't know. So I sez it should open in. This is the way it works out: 'Place yourself in there. The door openin' in, say about forty-five degree. This gives you air and lets the sun beat in. Now, if you hear anybody comin', you can give it a quick shove with your foot and there you are. But if she swings out, where are you? You can't run the risk of havin' her open for air or sun, because if anyone comes, you can't get up off that seat, reach way around and grab 'er without gettin' caught, now can you?'

Sure enough, one way of telling that the little building you

Bob Poplett of Peacehaven with a typical privy bucket.

come across was built as a privy is the swing of the door. It's a detail, we might note in passing, that detracts from the suitability of privies as garden sheds, but most of us simply put up with the inconvenience of trying to manoeuvre lawnmowers out of a tight space with a door in the way.

A second clue is the pattern of cut-outs, either within the door itself or, more commonly, along the top and bottom:

'Now, about ventilators, or the designs I cut in the doors. I can give you stars, diamonds, or crescents – there ain't much choice – all give good service. A lot of people like stars, because they throw a ragged shadder . . .'

Sussex privies rarely display anything more elaborate than a row of v-shapes to let the air in and out, with an occasional small window high in the wall above the eye-line.

When it comes to the business part of the building, it becomes obvious that our Lem is thinking of the most basic, the bog-standard privy:

'Now, about the diggin' of her. You can't be too careful about that,' I sez; 'dig her deep and dig her wide. It's a mighty sight better to have a little privy over a big hole than a big privy over a little hole. Another thing; when you dig her deep you've got 'er dug; and you ain't got that disconcertin' thought stealin' over you that sooner or later you'll have to dig again.'

What we have here, in other words, is a seat built over an open pit. Most of the early Sussex privies would have been like this, with an opening at the back to allow a shovelling out once or twice a year – or as infrequently as necessary. Although the pit might be covered with boards, Fred Osborne's story shows that this wasn't always the case.

'Until the 1914–18 war we lived in an old cottage – father, mother and four boys. The loo was approximately a hundred

Lem Putt offered stars, diamonds and crescents, but the triangular cut-outs at the top of this door in Hurstpierpoint are the standard Sussex ventilator design.

A 'pigeon hole' ventilator design at Beddingham. This little privy survived the explosion of a German bomb about twenty yards away during the Second World War.

The late Miss 'Ray' Raymonde-Hawkins, founder of the Raystede Animal Sanctuary near Ringmer, stands at the rear of an ancient privy in her grounds. A modern drain cover now prevents a nasty fall into what was once an open pit.

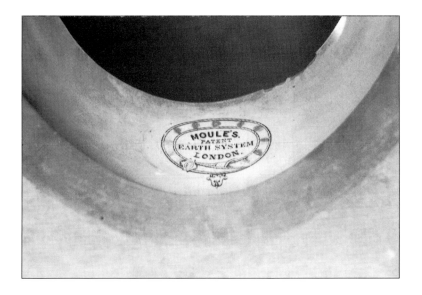

Heaven and earth. Henry Moule's calling was the church, but he's best known for his earth closet. This fine survivor sits in a garden pavilion at Tanners Manor, near Heathfield. A tug of the handle released a quantity of sifted earth on to the raw material below.

yards up the garden. It was close to the trunk of a huge yew tree, under which was what could only be described as a bog, because when the hole under the seat began to fill up father would flush it out under the tree – a large saucer-like hole. The yew tree spread out over this area, but there was never a smell.

'We used to play a lot with bows and arrows, and have fights with them. Our mother was worried about this, and if she could get hold of the bows they would be chopped up, so we used to climb up the yew tree and hide them.

'Another trick was to climb to the top of the tree and then jump out onto the spreading branches over the bog below. The branches would bend under our weight, lying one on top of the other. We would then climb up the sloping branches and back down the trunk.

'Do you wonder mother got worried, and can you imagine what we'd have got if she'd caught us? If we had slipped through the branches into the bog below, well – I have no idea. We didn't know how deep it was, and it was always wet.'

Truly disgusting, and a reminder that when we come across an old privy with a bucket under the seat we are actually witnessing a sign of progress in down-the-garden sanitation. Emptying, of course, needed to be a much more frequent exercise (and what happened to the raw sewage is another matter), but so-called 'bucket and chuck-it' at least got the stuff away from the building and gave the users, as it were, a clean sheet.

Some privies had a lid over the hole, but there was no getting away from the fact that, once seated, you were poised above an open can of urine and excrement. The best that could be done was to throw something down on top of it to reduce the smell, and various inventors tackled this challenge over the years.

The Reverend Henry Moule's patented 'pull-up' earth closet of 1860 was among the most successful of these, and I was delighted to be invited by Joan Baker to inspect a fine specimen in her garden at Tanners Manor in Lions Green, near Heathfield – a house originally built for the Fuller family of iron-making (and folly building) fame.

How could I resist a letter which told me the Moule privy 'was incorporated as part of a lovely pavilion, set beside a lake and within a garden possibly designed by Gertrude Jekyll or William Robinson'? As my photographs show, I wasn't to be disappointed. (The garden is wonderful, too!) It's a simple device which dropped dry earth, charcoal or ashes on the waste with a pull on the handle. The end product was apparently both sterile and odourless.

The water closet was, of course, the cleanest answer to the problem, and the Victorians devised some interesting portable indoor versions for people of means: porcelain commodes with leather seats inside carved mahogany cases, and pumps to activate small lavatory tanks hidden away inside.

Michael Ridley's father decided to get up to date with water in his garden privy, although the luxurious trappings were markedly absent.

'In about 1980 my parents took on the task of renovating an old house in Ashburnham, near Battle. It was the house that my mother had been brought up in. It was the usual story of no bathroom or indoors loo, relying on a tin bath in the sitting room and a toilet up the garden.

'This toilet was a brick-built affair, originally housing a bucket style toilet, but this was soon replaced, by my father, with a pseudo flush toilet. The "flush" was provided by a bucket of water that was stationed just outside the loo, and which had to be poured in manually. Within a few years a small hand-dug swimming pool provided the water supply – still via a bucket.'

Where the spoil ended up we aren't told, but presumably Michael's father was returning to the old pit style of disposal. In the towns, as we have heard, the same method would have been used, but with the privy sitting over a drain into the sewer.

Michael Charman recalled an interesting device attached to a loo connected to the water supply.

An ingenious self-contained model in Bert Winborne's collection. The date is somewhere about 1894. 'It's got a nice ashwood seat attached to an iron frame, which you lift up by means of a round knob handle. It's easy to empty. As you lift the lid up the container inside tips up and shoots everything away. There would have been a trench at the back.'

'I started my career in psychiatric nursing at St Francis Hospital, Haywards Heath, in 1956. When I was first on the wards, many of them had "self-acting" toilets. The toilet was encased in a wooden box and the seat (square with the usual oval hole) was held up at an angle of 45 degrees by a spring mechanism.

'The user sat on the seat and, upon rising, the seat rose up, thus flushing the toilet. This was because patients often didn't manually flush it. I believe this to be a Victorian invention. One had to sit centrally on the seat to avoid pinching one's flesh as the seat went down.'

Ah, bottoms! Lem Putt is, as we might expect, expansive on the subject, but on this occasion he proves unhelpful. Having built 'the average eight family three-holer' for a farmer, he is called back to discover why the hired hands are wasting up to an hour of their employer's time on each visit to the privy:

'I looks at the seats proper and I see what the trouble was. I had made them holes too durn comfortable. So I gets out a scroll saw and cuts 'em square with hard edges. Then I go back and takes up my position as before – me here, the Baldwins here, and the privy there. And I watched them hired hands goin' in and out for nearly two hours; and not one of them was stayin' more than four minutes.'

Comfort was, of course, the usual aim, and I have had numerous glowing reports of well-fitting wooden seats which were a warm pleasure to nestle into. Let one (from R.S.L., a semi-anonymous contributor from Sompting) suffice for them all:

'The seat in our first privy was a round hole cut in a piece of rough hardwood, so rough that you were frightened to slide off too quickly in case you acquired a splinter in a tender area. Also the hole was so far back and so high that my little legs couldn't reach the ground, so sliding was inevitable.

'In late 1942, however, we moved right into the village, into a cottage that had running water, although unfortunately we still had oil lamps, candles and a bucket privy just 25 feet (rather than a hundred, as previously) from the back door. But this time the lav seat was utter luxury, because it was made of thick soft wood. Being about three inches thick and of natural pine, it was beautifully smooth and shaped to hold, and almost caress, one's bottom. And in winter it had a warmth of its own.'

Wood can warp, however, as old Jim Copper memorably testified:

'The dunnicks were built in pairs set back to back under one roof and with a couple of pieces of 4" by 2" timber passing through the dividing wall to carry the wooden seating with the necessary holes over the buckets. Over the years these timbers had worn at the point where they passed through the wall, and there was a certain amount of play, which resulted in the seats moving up and down like a see-saw with the wall as a fulcrum.

'If there was no one from next-door in residence the seat would jolt down an inch or two as you sat on it, and if anyone turned up while you were there you could tell pretty much who it was by their weight. I could hold my own with either of the two girls, and if I sat down real hard like I could just about break even with their mother. Although there would be a certain amount of up and down work as she shifted about.

'But when the old man took over I would shoot a couple of inches up in the air, and remain there until he had finished. And you had to be mighty careful not to have your fingers under the seat when he got up, or you would pinch them on the rim of the bucket.'

Fashioning 'comfort stations' that would do for everyone presented something of a problem, however.

'My uncle Tom was a carpenter,' Dot Burrill told me, 'and he made his privy to suit himself. He was a big man, so the hole he made for his own bottom was very large.

'Unfortunately the housekeeper was a rather small woman,

and on one dreadful occasion she slipped down into the hole and got stuck. She had to be hauled up out of it.'

John Christie hit on a more or less successful solution (we are here talking about something grander than a garden privy) when he built his opera house at Glyndebourne. We find a reference to it in the Lyttelton, Hart-Davis letters, George Lyttelton writing the following on 14 July 1960:

'I hope Ruth enjoyed Glyndebourne. I wonder if she met John Christie. He is a very Pickwickian figure now, physically as well as mentally, which he always was. He took enormous trouble over everything at Glyndebourne down to the smallest details.

'There is something engaging about what we would call the Prussian thoroughness, but for the sly fun in it, that in the cast's lavatories the ladies' seats are of different sizes because, as JC simply explains, the rumps of the prima donnas vary greatly in size. He once gravely admitted that the shapes as well as sizes varied, but regretfully decided that it was too delicate a problem to tackle.'

Multiple holes were fairly common up the garden path, often with a large hole for the adults and a smaller one (lower down, or with a step up) for children. Why people installed three- and four-holers it is difficult to say. I have been solemnly assured that the grown-ups would use one for solid matter and the other for liquid, so that the latter – with its high ammonia content – could be used as a 'mordant' for fixing dyes in wool. This makes sense in woollen areas, perhaps, but surely has no relevance to the average Sussex privy.

The obvious answer is that people sat side by side to go about their business, as soldiers or hop-pickers would, but not one of the people I heard from admitted to sharing the place with another member of the family, children apart. Perhaps some folk had a 'his and hers' arrangement, so that each gradually took the imprint of a particular bottom.

Joan Lacey had an unsettling fancy when she visited her aunt and uncle at Holt's Cottages on the Balcombe road.

'While my family were talking I would wander up to their privy – several steps up from the back door. A sort of house, as I remember. It had very clean scrubbed wood and buckets underneath, but it had two holes side by side.

'As a child I couldn't imagine what it would be like if someone came in to sit beside me – but it didn't happen.'

Elizabeth Skinner of Worthing recalled a rite of passage denied to the youngsters of today.

'I certainly remember our two-seater,' she enthused. 'It was a proud day when I graduated to the grown-up seat, complete with catalogues from J. D. Williams hanging on the wall. No Andrex in those days!'

[6]

BUM PAPER

'Newspaper was cut into 8-inch squares, with string
threaded through one corner. When you think what
newsprint was like then, there must have been
many rather grubby bottoms.'

Paper of all kinds is so abundant today that it is hard to imagine
a time when there was a need to wipe your backside with anything else. The 18th-century Earl of Chesterfield clearly believed
in it, though we are hardly likely to follow the advice he gave his
son and tear out pages from a book of poetry ('It will make any
book which you shall read in that manner very present to your
mind'). Squares of lavatory paper were recommended as a pre-
ventative against piles in 1857, and the British Patent Perforated
Paper Company introduced the first toilet roll in 1880, but paper
has been the commonplace material of the privy only since the
1930s.

Grass and leaves had always been freely available to those
stranded in the countryside – the dock leaf being especially
prized for its strength and flexibility – while for centuries the
less rustic alternative was a bundle of rags which could be
washed and used again. But then, almost overnight, the spread
of cheap and popular newspapers presented a wonderful two-in-
one gift to those keeping their lonely garden vigil: they could
read when in need!

'I liked to read while I was in there,' Hazel Burchell told me. 'I
think I got that habit because there was always newspaper available, as well as the dreaded Bronco.'

Children found themselves given the task of cutting up the
'daily' into neat squares and threading them on a string so that
mum or dad could hang them on that nail in the wall.

The *Daily Mail* was more friendly than most . . .

A rare, time-worn survivor of days gone by in a Sussex privy. The wording on this rusting loo roll holder reads: FAMILY TOILET REQUISITE CO. LONDON.

Not that the 8-inch square was universal. When the British Engineerium at Hove staged a lavatory exhibition a staggering 55,000 people came through the doors, and several recalled a niggardliness (no doubt caused by poverty) in the cutting of the paper.

'One preferred size was 4" by 2", says the Engineerium's curator, Jonathan Minns. 'When you think about it, the dimensions were similar to a dock leaf. You could hold it between finger and thumb, and draw it across.'

There were alternatives to newsprint.

'When we lived at Herstmonceux', said Nibs, 'we had a massive stone to sit on the paper and keep it flat. We used cut up comics, which gave us something to read.'

Doubtless the printing ink did often leave its traces (though I have heard that the *Daily Mail* was more friendly than most in this respect, and posters were reckoned to be both strong and run-resistant), but only the fastidious make a close study of their own rear ends, and there are usually few others entitled to share the view.

Bob Poplett had an early taste of luxury when he was growing up in Peacehaven.

'My dad used to give me and another boy a job in his fruit shop, pulling off the orange, lemon and apple papers. All imported fruit from abroad was wrapped individually in paper about 9 inches square, slightly greaseproof to stop cross-contamination of any disease while the fruit was getting here from abroad. In those days, of course, ships were slower and journeys were longer, and you had the risk of a whole cargo going bad.

'We used to take all these pieces of paper – about a hundred – put a nail through the corner about two inches in, then thread a bit of string through and sell them for people to hang up in their lavatories.

'Bum paper. Lovely smell to your bum!'

Gardeners, too, could count on a better class of wipe.

'We used the *Telegraph* and *Times*,' said Bert Winborne, the ancient sage of my book *The Upstart Gardener*. 'There was nothing else. It was thrown away from the big house.'

But those old gardeners were often set to work some distance from any privy with its minor comforts, and then Bert learned the joys of his fellow countrymen through the ages.

'When you were helping in the farm, and in the winter time you worked out, you might be working two or three fields away from the farm. Someone would say: "I want a crap, mate." You'd tell him "Hedgerower".

'They used to sit down and wipe their backside with a lump of grass, see. I used to do it, too. It's better than paper. We used to get the end of the old swaphook, stick it in the ground, cut it round, lift it up, put the old muck in there and put the turf back on top.

'Wet grass is very refreshing.'

It was Michael Charman who told me about what was probably the most expensive piece of loo paper ever used in Sussex.

'In the early 1950s,' he said, 'we were living in a farm cottage in Wivelsfield Green. Consequently there was a brick privy in the garden, with a bucket which we emptied into a hole in the corner of the garden. At this time we didn't have toilet liquid for sanitation.

'One day a family friend visited, and in the course of time she went outside. On her return she began looking for a £5 note which she'd had in her hand before she went.

'We all searched for a long time – and then came to the dreadful conclusion that she must have used it in the privy.'

[7]

A CHAPTER OF ACCIDENTS

'It is the opinion of the medical attendant that asphyxia by the sulphuretted hydrogen gas caused instantaneous death.'

Of all the many accidents that have befallen privy users over the years, one of the most dreadful occurred in Lewes on an April evening in 1856. The story, with the above verdict, was reported in *The Times* on 1 May:

APPALLING DEATH. Considerable excitement was caused in the town of Lewes on Thursday morning by a rumour that a young man named Matthew Gladman had met with his death by falling into the soil of a water closet the evening before, on the premises of Dr Smythe, High Street. It turned out to be too true. It appears that the poor young fellow incautiously went to the watercloset in the dark. He found that the door was fastened with a rope, the boards having been taken up preparatory to the cesspit in connexion with the closet being cleaned out. That rope he appears to have untied, and to have tumbled at once into the soil below. The body, on being extricated, was wiped dry, and galvanism applied, but in vain.

Galvanism, in vogue at the time, involved passing a small chemically-induced electrical charge through the body.

A less detailed account of a similar gruesome accident can be found in the Westham burial register for 14 December 1707:

William son of William & Sarah Weller aged 14 years, who dyed by a fall out of a Garrett into a Closett underneath.

Most other lavatory misfortunes have their amusing side – at

DRURY-LANE—ENGLISH OPERA.
Under the Direction of J. H. Tully and F. Kingsbury.
THIS EVENING, FRA DIAVOLO.
Messrs. Henry Haigh, Durand, Manvers, Bernard; Miss Dyer and Miss Fanny Reeves.
To conclude with Mr. and Mrs. W. J. Florence's impersonations in
THE YANKEE HOUSEKEEPER.

THEATRE ROYAL, HAYMARKET.
Under the Management of Mr. Buckstone.
THIS EVENING, THE EVIL GENIUS.
By Messrs. Buckstone, Compton; Misses Reynolds, Swanborough, &c.
After which, Perea Nena and the Spanish Dancers.
To be followed by GRIMSHAW, BAGSHAW, AND BRADSHAW.
To conclude with THE POSTMAN'S KNOCK.

SINGULAR ACCIDENT.—A singular accident happened to the Simla steamer, which went to the naval review last week. As she passed the station of the middle buoy in the Solent the buoy was found to be missing. The Simla arrived off Southampton without anything apparently the matter with her; but while at anchor she did not swing with the tide. Soundings were made, and it was found there was plenty of water to enable her to swing. A diver was sent down to examine her, when he found the chain of the middle buoy coiled round her screw. She is unable to proceed to the Crimea until she is docked in consequence of this accident.

APPALLING DEATH.—Considerable excitement was caused in the town of Lewes on Thursday morning by a rumour that a young man named Matthew Gladman had met with his death by falling into the soil of a watercloset, the evening before, on the premises of Dr. Smythe, High-street. It turned out to be too true. It appears that the poor young fellow incautiously went to the watercloset in the dark. He found that the door was fastened with a rope, the boards having been taken up preparatory to the cesspit in connexion with the closet being cleaned out. That rope he appears to have untied, and to have tumbled at once into the soil below. The body, on being extricated, was wiped dry, and galvanism applied, but in vain. It is the opinion of the medical attendant that asphyxia by the least in retrospect and certainly for those not directly involved – and we shall pass swiftly from tragedy to slapstick. Here is Michael Charman at St Francis Hospital:

'One ward at the hospital was on the ground floor, but had toilets beneath with access from outside so that patients didn't have to enter the ward in order to go. When the urinal became blocked I phoned for the plumber, who removed the trap and threw down a bucketful of water.

'Unbeknown to me, someone else had also called out a plumber – who had just removed the downstairs trap, and was looking up the drainpipe as my man threw the water down!'

It was all too easy to lose things down the hole, as Jean Lawrence discovered. At least her little place in Lewes was above an accessible drain.

'At one time my husband dropped his watch down the privy, so the manhole cover came up while I threw down water. It came along, and he retrieved it. Good job it was waterproof. We were so poor he couldn't afford a new one.'

A story from Irene Squires' family history also highlights the financial implications of losing something that today, in the circumstances, might well be thrown away.

'Emily from Litlington and her four siblings had to walk to school at Alfriston every day. They would either go across the fields (subject to tides and weather) or via the road and a long bridge: the Cuckmere had to be crossed somewhere along the line.

'Three-year-old Ernest had to tag along and he of course got very tired on the way home, so had to have a piggy back. My mother stood him in a "bog" one day so that he could climb on to her back more easily, and the lid just happened to be rot, and in he fell.

'Between them they fished uncle out and carried him home, complete with plenty of attachments – and all they had was a well in the main street – but with only one shoe. Calamity! Grandma was furious, and sent mother back to the privy to fish out his missing shoe. They only had one pair each.

'Oh, the joys of country living at the turn of the century.'

Myrtle Hancock told a similar story, but from a personal point of view.

'When I was about four years old I was playing with my great friend and neighbour, and we discussed whether it was safe to walk on the lump at the back of their loo – and I was foolish enough to try it, with disastrous results.

'I may say that my mother was not at all pleased by the high-smelling daughter who came crying home, though my elder sister set up a bath of water and cleaned and consoled me.'

But the vilest story I heard in all my travels and correspondence (the faint-hearted should not read this before sitting down to eat) was told to me by Mary Shelton.

'I can remember one Christmas Day we had an aunt and uncle come for Christmas dinner. Dad and Uncle John walked off to the local pub a mile away for a pre-dinner drink, while Aunt Alice stayed to help mum.

'The dinner was ready to dish up at 12 o'clock, but there was no sign of dad or uncle. Mum was just getting worked up when dad walked in. "Where have you been?" she demanded. "And where's John?"

'"He'll be here in a minute," dad said. "He's just gone to the bog."

'It was another ten minutes before uncle came in, looking, as mum said later, pretty sheepish.

'Aunt Alice never did know what had happened, but mum and I did. Dad told us after they'd gone.

'Uncle had had a bit too much to drink at the pub, and he'd just managed to reach our bog before he started vomiting. Unfortunately he lost his false teeth as well.

'Need I go on?

'There they'd been all that time, desperately trying to retrieve them so that he could eat his Christmas dinner without Aunt Alice knowing anything about it.'

Someone who seems to have been *asking* for a nasty accident to happen was one of Myrtle Hancock's relatives.

'During the war,' she told me, 'when we had an air-raid warning, my cousin's wife, who was staying with us after being bombed out of Portsmouth, used to take my baby and her two little girls aged three and two and shelter in the loo, because she figured it was a very small target.'

This curious logic would surely have been shaken had the lady had a word with Gladys Fraser of Handcross.

'My father, Harry Pateman, was the gamekeeper at High Beeches,' she told me. 'We had a house there, with a privy at the bottom of the garden. It was a little wooden structure, single-holer.

'In January 1942, a landmine came down during the early hours, missed the house but demolished the privy. Father had to rebuild it.

'One evening the following August I went down there to use it, and I'd just got back into the house when we heard a doodle-bug coming down. It missed the house, landed next to the privy and blew it to smithereens!'

An amazing coincidence, although we have perhaps already moved on from genuine accidents to foul intent.

[8]

DIRTY TRICKS

'The village school had earth closets which were fitted
with trap doors at the back for emptying. The boys would
pick long stems of stinging nettles at playtime, creep
round behind the girls' privies, open the trap doors
and push the nettles up to sting the girls' bottoms.'

Mary Shelton's anecdote has something in common with most of
the stories I was told of privy mischief: the culprits were boys.

'The younger ones were all made scared of the dark by the
older boys,' remembered Fred Osborne. 'The loo was about a
hundred yards up the garden, surrounded by trees, so it got
dark early in the evening.

'When it was necessary, another boy would go up with you.
Often, before you had finished, the other one would bolt off
home, leaving you alone. No window to the loo, only the open
door. No lights visible. Total darkness. You were scared stiff.'

Pauline Lavington recalled one particularly sticky moment.

'On one occasion my brother was grounded for some trouble
he'd been up to, and he decided to get his revenge. He painted
the seat of the privy bright green with red spots.

'It was gloomy in there, so you couldn't see very well, and I sat
on the seat and had the paint all over my backside.'

The darkness certainly helped an imaginative gang of four at
West Chiltington, as Ella Lipscombe related.

'My father, his brother and two cousins lifted the sheet of ice
off the water butt one winter's night. They put it on the privy
seat and waited behind the bushes for their grandfather to come
home from the pub.

'Needless to say, they thoroughly enjoyed the yell when
grandfather sat on the ice.'

This privy is in the garden of the former toll cottage from Upper Beeding, now preserved at the Weald and Downland Open Air Museum, Singleton.

If a girl is officially the villain of our next tale, Dot Collyer could certainly put part of the blame on her brother.

'He had an air gun, and one day he asked me to have a try. Let's just say the pellet went a bit astray.

'Someone was sitting on the loo at the time, and the pellet hit the galvanised roof, making him jump. Out he came, trousers round his ankles, looking none too pleased. We scarpered, but we were told off pretty severely later on.'

Doreen Kallman adduces sisterly provocation as mitigating circumstances for her privy mischief when they lived at Old Rectory Cottage in Westmeston after the last war.

'My older brother and sister always reckoned I could get away with anything, but my parents got very cross about the revenge I took on my sister.

'She had a pretty bad temper, and when she was in that mood she would go and lock herself in the privy. That meant, of course, that we were jumping up and down outside waiting to go.

'What I'd do was go inside at the first opportunity, bolt the door and then climb out of the small window high in the wall. I was seven or eight at the time, so I could just squeeze through.

'It was her turn to do the jumping about, then, while I had my freedom outside – though I had to climb back in again and unbolt the door pretty quickly once my parents heard about it.'

[9]

Yer Own Back

Foule privies are now to be cleansed and fide.
Let night be appointed such baggage to hide,
Which buried in garden, in trenches a-low
Shall make very many things better to grow.
Thomas Tusser (1580)

Doing your business away from the house was all very well, but disposing of the stuff afterwards was an unpleasant task. Perhaps it was retribution for all that skulduggery they had perpetrated in their younger days, but this was undoubtedly Man's Work.

Boys too, however, sometimes had to play their part.

'When I was a kid at my grandfather's place at Rye Harbour,' said Bob Poplett, 'most people still had dirt closets. They used to save their ashes for that off the fire, and put it in a heap. They'd get a bucket and say "Go on, boy, take it out there." You shook it on the heap of evil-smelling stuff and it sort of masked the smell, kept the flies off it and consolidated it.'

In the town there was often a take-away service for those who paid their rates.

'You got men coming round between 11 o'clock and one or two in the morning – that's a Saturday night to Sunday morning – and they'd have the job of getting buckets or a wheelbarrow and carting away the contents of the privy.

'Often they were built on the side of a bank, and you got at it from behind. You had to get down in the lower area so you got under the seat, and you got a shovel in it. And you took it out the front and tipped it on the road outside in a heap, and then the men would come along with an old tip cart about half an hour later through the whole street, and the next morning you'd see half a bucket of powdered lime spread all over the stuff. It would burn away the disease, but it would stay there all week. People walked through it.

'I can remember the last lot in the area where I live now. It was in New Road, Newhaven. There were about ten cottages in a row – they're still there now. It was after the last war, and I wondered what all the white patches were.

'They were the only houses in Newhaven that weren't on the main drainage. They tipped the muck out the front, on the road, and they put lime on top afterwards, but the white patch was there for days, especially if it didn't rain. That was until about 1950.'

Bob Copper lived not far away at Rottingdean, and – as he wrote in his wonderful autobiography *Early to Rise* – he helped his father empty cesspits:

He had mounted a heavy, iron hand-pump on a stout, wooden, trestle-type frame with two handles at each end by which it could be carried in the manner of a stretcher. To the pump were attached two lengths of four-inch, armoured, rubber hose, which some years before had been used for filling the water-tank of the steam traction engine on the farm, and with this contraption we used to empty cesspits at five shillings a time.

As our methods were a little unorthodox – merely running the outlet pipe over the garden fence or through the hedge and pumping the effluent out on to the nearest piece of spare ground – we usually operated after dark.

When I made a radio programme about the development of neighbouring Peacehaven, I discovered from Doris Leach that standards of hygiene were similar there in the 1920s and 30s:

'My father knocked bricks out of cesspits. He'd wait till dark – ten o'clock at night – and he'd empty so much, climb down inside and knock out a brick, or several bricks, so that instead of having to have it emptied every two or three months or whatever, it would seep through the ground and not fill up. It was very illegal, and I think he got paid ten shillings for that. And he would come home and the whole house would absolutely stink!'

The contents of the cesspit provided a rich fertiliser for gardeners in the days before today's magical compounds had been invented, as Thomas Tusser's verse atop this chapter testifies. ('Fide', incidentally, means 'purified'.) In *The Upstart Gardener*, Bert Winborne described an unpleasant operation:

'We'd get a builder in to take the big slab off the old cesspit, and we'd dig the sludge from that into the garden. It used to be done on a moonlit night when it was frosty and that helped to keep the smell down.

'There was a crust about a foot deep and we cut it out and took the pieces up into the shed to dry out. Then we'd beat them up with a big old wooden beater and that was shovelled again into wheelbarrows and taken to the potting shed and put under the bench in a big bin.

'We had a name for it. It was called Yer Own Back. The head gardener knew the best mixtures to use for everything, and he'd tell you so much leaf soil (there was no peat in those days) and how many handfuls of Own Back if you were potting stuff on. It wasn't a joke. That's what it was called.

'The early peas liked it, too. We used to grow them in the greenhouse, and we'd give them a shake of this dry sludge each side of them and draw it up with a broad hoe, and that used to give them a bit of colouring. By Jove, we used to get some peas!'

Handling the crust can't, however, have been the worst job.

'Once that was off we'd have long scoops and bodges – tanks on wheels – and we used to fill them up, run them round the garden and tip them up on the ground. Then we'd lime everywhere to keep the pong down.

'It certainly put plenty of lime and iron in the ground and, you know, we could grow just about anything on that old sludge.'

Even the simplest call of nature was put to good effect. There was a large copper behind the woodshed, and there the careful

He who sups with the Devil . . . Bert Winborne poses with a large 'bodge' and a scoop used for emptying cesspits.

waste-not, want-not gardeners would repair.

'The arum lilies liked that,' Bert said, 'and just a little of it would bring the leaf veins out on a fern.'

There was, it seems, no special name for this natural fertiliser.

'The head gardener would just ask us to take the bucket round the side of the laurel bush. We had an iron saucepan, and we'd ladle it out and mix it with water.

'But you had to be careful when you were caught short. Her ladyship might be out with a big basket, cutting flowers, and if she heard this sprinkling sound like rain she'd whip her head round and you'd be in real trouble!'

The contents of the cesspit may have been some considerable time forming their solid crust, but many a keen gardener simply dug the few-days-old contents of his privy bucket straight into the soil.

'It had to be emptied in a specified piece of the garden,' said John Goddard, remembering the days when he helped his father at Milton Street. 'And he didn't want much water in this bucket, first because if there *was* it had to be emptied frequently and, secondly, because too much didn't do the garden very much good either.

'This bucket had to be carried out of the yard and into the garden outside. When we got old enough we had to help with the emptying. It was a special shaped bucket which in later years was used as a coal bucket.'

Although privies above mains drainage usually made no uncomfortable disposal demands, Norma Noakes recalled the occasion when a blockage lifted the inspection cover next to her grandparents' privy in Ladysmith Road, Brighton.

'My grandfather had to dig it out and put it in buckets, and he then threw it all over the garden. I remember that it dried like a skin.'

Olive Morgan wondered, as we all might, about the health risks involved.

'Naturally the "dead" had to be buried in the garden in deep holes. We had wonderful fruit and vegetables. But I've often puzzled about the council in Eastbourne refusing to let the estate have bullocks on the hills, their argument being that they'd pollute the water for the town. What about all the "dead"?'

For some, there was help available.

'Up to about 1930,' remembered Myrtle Hancock, 'inside ours was a deepish hole, and the privy was open at the rear. Each summer this was emptied by a man with a horse and cart using a long-handled tool with a bucket-like end. It was carried away and conveyed to the dung lump in the farm yard – a summer task, because it would hopefully be comparatively dry.'

But it's the wet that R.S.L. recalled.

'We lived in the heart of rural Sussex, and as I grew older I had to take my turn at emptying the very large bucket that slid under the seat.

'First one had to find a site in the large garden – one that hadn't been used recently or, better still, one that hadn't been used at all. The idea was to walk around the vacant vegetable patch until you found yourself standing on reasonably hard ground. Then you dug a hole at least three feet deep.

'The trouble was that sometimes my brothers would rush the job and dig only about twelve inches down before emptying the bucket, and this was where the soil would be very soft and sticky under foot. If you found yourself standing on soft, squelchy ground – forget it.

'Of course, this bucket job had to be done during darkness and with only the aid of a battery torch. Presumably the smell didn't travel too far after dark! My biggest problem was my height, or lack of it. Being only a young lad – I was about ten – carrying a bucket full to the brim was almost beyond my capabilities.

'With my short arms and short legs, I walked very slowly while concentrating on trying to keep the bucket off the ground, but frequently I'd bump, and slop the contents down

Great for emptying. Ken Morris of Hurstpierpoint remembers this bucket, with its generous spout, being used in a caravan site privy when he was a boy.

one leg. Then, when I returned indoors, I'd get shouted at for making a smell.'

Ron White has lived in his house alongside the A27 at Beddingham since he was four years old, and he used the privy which still stands there until the late 1950s when he was approaching forty.

'There was a large pit at the rear of the privy, covered I seem to remember with wooden sleepers. It was later modernised with a bucket. Dad emptied the pit once or twice a year and filled the

This is British Telecom's 'Bowl GI 9in.', described as follows: 'Galvanised iron 9in. diameter bowl $3\frac{1}{4}$in. deep with a 5in. wooden handle. Used for baling water out of joint boxes or warming water on a propane stove to assist the mixing of resin.'

Very useful – but why is it routinely issued in Sussex to crews who do neither of those things?

Well, just think what it's like for a chap working for hours at a time in an urban or suburban street, far from a convenient field or copse, especially if his bladder is on the weak side. Where can he go? The next time you see a Telecom man emerge from his van and tip a ladleful of liquid down a convenient drain you'll know what he's been up to.

runner bean trench – but not at a time when the buses would be going by!'

Hazel Burchell, thanks to her gender, had only to watch rather than participate.

'I remember my grandad used to dispose of the bucket contents. All around the edge of our orchard was a continuing deep trench in which the "soil" was deposited and then back-filled with earth.

'A neighbour of ours had a quite different system, rather like a cesspool. Above ground there was some sort of pump arrangement, and he used to draw off the liquid for his vegetable garden. The smell was appalling – enough to put anyone off veggies for the rest of their life!'

Which, in some cases, it did.

'My grandfather at Houndean Farm, on the outskirts of Lewes, used to dig out the privy once a week,' Diana Howell told me, 'and he'd feed the rhubarb with it.

'My aunt eventually discovered where it went, and she wouldn't eat rhubarb ever again – wherever it came from.'

[1 0]

TALL TALES — AND TRUE

> An elderly monk from Teglease
> Said 'Bury me where I drop, please.'
> As soon as he spoke
> His privy seat broke:
> He's interred in his chapel of ease.

Our fascination with an act which we usually try not to acknowledge has given rise to a host of privy stories. I suspect that the four which follow – all told to me while researching this book – have many subtle variants, sometimes incorporating the names of local people known to the teller and his audience.

My thanks for this selection (with apologies for anything lost in the retelling) to Norman Edwards, Peter Bailey, Bob Copper and Consuelo Lerwill:

An ancient Sussex couple have reached the stage where they can no longer look after themselves properly at home, so they decide to sell the cottage they've inhabited since their marriage and move into residential care.

A rather well-spoken lady looks over the property, and eventually asks, with a blush, if she may inspect 'the usual'. She is directed along the garden path.

On her return, they notice a shocked look on her face.

'My man,' she cries, 'there's no lock on the door!'

He shakes his head sympathetically.

'Don't you be troublin' yourself, missus,' he says. 'We bin 'ere all of fifty year, and no bugger stole a bucket o' the stuff yet.'

An old man and his wife are having a drink at a Sussex inn when he gets taken short. The trouble is that he's got two wooden legs, so finds it hard to get about. His wife watches him hobble off down the path, and after twenty minutes or so she starts getting a bit concerned.

She sets off for the privy, and she's getting close to it when she hears her husband's cries. He's stuck halfway down the hole with his legs in the air.

'Help!' she shouts to a young man who's passing. 'My husband's trapped in the closet.'

'Can't be,' he replies. 'I just slammed the door on it 'cos someone left a ruddy wheelbarrow in there!'

A caller trying to trace the occupant of an isolated cottage in a wood deep in the heart of Sussex gets hopelessly lost. He scrambles about, making his way through a dense tangle of brambles and undergrowth, until at last he finds himself at the door of a small flint building.

He gives it three hefty thumps with his fist, and waits. After a while the door slowly opens a few inches, and the face of a small boy appears, a little above knee-high.

'Hullo, sonny,' says the man. 'Is your father in?'

'No, sir,' comes the reply, ''e went out when mum came in.'

'Well, then, is your mother in?'

'No, sir, she went out when my sister came in.'

'Yes, yes,' says the caller, rapidly losing patience. 'Then I'll talk to your sister.'

'You can't, sir, 'cos she went out when I came in.'

'Oh, my God,' says the man. 'What is this, a mad-house?'

'No, sir, it's a shit-house – the cottage is up the end of the garden path!'

Young Fred goes off to be a soldier, and when he returns from the war he can't believe how backward things seem at the Sussex farm where he was raised.

'It's time for change!' he urges his father, pointing at the old carthorse and the antique plough. 'Move with the times.'

He runs a scathing eye over the delapidated privy, pulls a hand grenade from his pocket and hurls it at the old brick-and-tile structure. There's an almighty explosion – after which his mother emerges, dishevelled and shaking.

'Must be something I ate,' she mumbles, staggering back indoors.

While all of these stories could, just, have happened, the best yarn of all those I picked while on my journeys turned out to be absolutely true. Roy Fuller's tale of an amazing coincidence also had a historical dimension which inspired some fascinating research.

'In 1978,' began Roy, a builder, 'I attended an auction at an old builders' yard at East Hoathly. This had been run for a great many years by the Hall family. During the 19th and for much of the 20th century they were very well known for first class craftsmanship, and many of the Victorian and Edwardian houses built in the district still bear the distinctive stamp of the Hall style and quality.

'However, as is often the case, the firm went into decline and eventually closed down. The builders' yard and buildings remained exactly as they had been left when work ceased. It was an amazing Victorian timewarp – the clutter on the benches and the tools hanging on the walls in the carpenter's shop; the paint shop with pestle and mortars, tubs of red and white lead, the tins of colours ready for mixing; the plumber's shop with the lead working tools; the treadle lathe, grindstones, stationary engines and saw benches; builders' wagons and handcarts, and even a covered saw pit. If only it could have been left intact it would have formed a fine museum.

'Everywhere you looked there were stacks of wood. The useful timber had been sold in the sale, but the remainder stood in every workshop and shed, and was either short ends or second-hand. I'd recently installed a large wood-burning stove and was always on the lookout for fuel. This was perfect, as it was dry and seasoned, and with the cheek of the devil I asked the auctioneer, Michael Clark, what they would be doing with it.

'I was told that it had no value, and that I could take as much as I liked. Next day, with several journeys in my truck, I removed a very large amount of it, which was soon stacked alongside my own stationary engine and sawbench.'

Now came a strange discovery.

'Among this debris was a very nice piece of mahogany which clearly had once been a privy seat, and which was far too good to burn. In stacking this in my workshop for possible future use, I realised that it hadn't always been a privy seat. On turning it over I discovered a set of Roman numerals carved on the back. At some time in the past these had obviously been decorated with gold leaf: the privy seat had been formed from half of a large clock face.'

And then the great coincidence.

'It stood in my shed for quite a time until a chance visit from my brother Peter, a plumber. He took one look at the privy seat/ half clock face and couldn't believe his eyes. In his shed, he exclaimed in amazement, he had the other half!

'He explained that long ago he'd been working on the modernisation of a house at the bottom of South Street in East Hoathly. An old privy had been stripped out for conversion into a modern WC, and the carpenter had set aside the fine mahogany seat. Peter had been rather intrigued by the numerals carved in the back, and so he'd taken it home.

'The two halves matched perfectly,' Roy added, 'and it was clear to us what must have happened. Halls must have removed this old clock face from somewhere, probably local, and were involved in fixing the replacement. Being careful, like all good tradesmen, they'd turned the quality timber over and re-used it in two of the houses they were building at the time.'

But where had the clock come from?

'We thought it might be the old clock face from East Hoathly church,' Roy said. 'That was square, and fixed diagonally with just a single hand, so it seemed the most likely candidate.

'When it was replaced by a new, up-to-date striking model there was apparently a rumpus among some of the gentry in the villages, because the striking of the hours would enable their workers to know accurately when they were entitled to down tools!

'We gave the two pieces to the then vicar, who said he'd probably put them in the church tower.'

That was the end of Roy's fascinating little tale, but it obviously left a few things hanging in the air. I visited the church, but could find no trace of the mahogany boards with their circular holes for bottoms.

The next step was to discover more about the history of that clock. Over to Eric Gould, who looks after the church fabric. He'd never seen the mahogany boards, he told me, and the current clock had been on the church tower since about 1875 – so if Roy's theory was correct, the sawn up timber must be well over a hundred years old.

'Could well be,' Roy insisted.

Eric did a little more digging. The previous clock had originally been fixed, not to the church but to the stable block at Halland Park, one of the houses of the Pelham family, Dukes of Newcastle and Earls of Chichester. But one of the dukes was a close friend of the Prince Regent, and he built Stanmer Park in order to be closer to the court.

Halland Park was demolished – and the clock was given to the church.

Could this really be what the Fullers had found? The next piece of detective work involved a visit to Clara's tea shop in East Hoathly, where Jane Seabrook has a marvellous collection of local history, including some old photographs. Had she by any chance (heart in mouth) an ancient picture of the church, clock and all?

Yes, she thought she had. And, while she looked, Jane told me

How are the mighty fallen ... The handsome clock on the tower of East Hoathly church was destined to serve more lowly purposes – as a pair of privy seats.

that the original clock had stopped working in about 1863, but that there had been no funds to replace it, so it remained in place for some years until a legacy from a parishioner paid for its replacement.

Now out came the photograph, and there was the venerable timepiece – square, with point uppermost, and with Roman numerals.

'That's it!' cried Roy. 'Unmistakable!'

What the former vicar did with the boards remains a mystery which may never be solved, but here we have one of the weirdest how-are-the-mighty-fallen stories ever told – the conversion of a lordly timepiece into a resting place for lowly bottoms.

[11]

WATER, WATER, EVERYWHERE

'There was great jubilation in the 1950s, as mum's
landlord was going to have a new toilet put in.
The rent was going up from 7s 6d to 10s per week.
What a let-down: a new toilet went in, but no cistern,
so it was still buckets.'

In 1875, at about the same time as that East Hoathly clock was being lowered from its tower to meet humbler requirements, Daniel Thomas Bostel of Brighton was proudly picking up a prestigious award for his Excelsior water closet.

It was the first of several. In the years to 1884 he exhibited the godsend he had designed and patented at the Norwich, Stafford, Eastbourne and London Health Exhibitions, and he won top awards at all of them.

Bostel hasn't left his name to posterity as some of his fellow (perhaps we should say rival) sanitary engineers have done, but in his book *Clean and Decent*, Lawrence Wright ranks him alongside Thomas Twyford and Frederick Humpherson – responsible, respectively, for the Deluge and the Beaufort WC – as foremost among those who have claim to be declared inventor of 'the washdown closet', the direct ancestor of what we use today.

'The washdown is simple and efficient,' Wright adds, 'and can be made in one piece. It has the minimum of exposed surface and the force of the flush passes through it unimpeded. Like earlier types, it may need a 3-gallon flush rather than the 2-gallon limit imposed by most water boards, and it can be shockingly noisy. But essentially the problem has been solved.'

Brighton stands as an excellent example of what the Victor-

Brighton's magnificent Victorian sewers not only serve the town today but are regularly open for guided tours. Here 12-year-old Gemma Clarke, a pupil of Tideway School in Newhaven, is escorted through the brick labyrinth.

ians did to tackle the vileness which had cost so many lives. A huge sewer designed by Sir John Hawkshaw was begun in 1871, seven miles long and so sturdy that it is still in use today – and explored regularly by curious and admiring subterranean visitors, especially during the Brighton Festival. Waste which had previously been discharged close to the shore from a number of separate outlets was now channelled into a single out-fall at Portobello, near Telscombe Cliffs.

But improvements in hygiene needed something else, too: water, and in much greater quantities than were sufficient merely to discharge waste matter down a toilet. It required a substantial flow to carry everything away to a (more or less) safe distance, and huge beam engines were installed at the Gold-stone Pumping Station in Hove, now the British Engineerium. They raised water from wells 165 ft deep in the chalk, and large

Did Queen Victoria, perhaps . . . ?

'One day about 1955,' recalled Bob Poplett of Peacehaven, 'I went to an auction in Middle Street, Brighton – F. T. Wilson & Sons, who'd been official contractors for the Corporation. Because the Corporation owned the Royal Pavilion they had the job of removing broken lavatory cisterns and putting new pans in. They took this old one out on one occasion, and the foreman said "You'd better put that up on the shelf, mate."

'Well, about 35 years later the firm went bankrupt, and I was in there and bought a lorry-load of stuff. Near the Hippodrome it was. I bought all the vitreous gear in one load, and the old foreman said "You'd better be careful with that, boy. It came from the Royal Pavilion."

'It's circa anything from 1850 upwards.'

amounts of it were used to flush the sewers. By the end of the century mortality rates in Brighton were down by a third.

All this healthy activity notwithstanding, flush lavatories arrived in many Sussex homes (and gardens) many years after the heady period of prestigious Victorian public works. Jean Lawrence's experience, quoted at the very beginning of this chapter, was by no means unusual: the 1950s, and still no WC.

Looking back, the Sussex people seem to have been a pretty stoical lot, accepting what crude comforts they had, and eventually rejoicing when something better came along.

'Father was wounded on the Somme,' Joan Prangnell told me, 'and when the war ended he had to seek work in the country.

'He came to Bodle Street Green in 1919, and their lodgings had a privy at the bottom of the garden with the chickens. In 1923 they moved into 1, School Cottages, where there was an

A typically neglected, if once modernised, privy – its rusting cistern draped with thick cobwebs.

The glory of the garden . . . The lack of growing space on the houseboats at Shoreham beach demands that a keen gardener be inventive and adaptable. (Photo courtesy of Chris Coates)

indoor toilet, and when my mother sat on it, she told my sister and me, she felt like a queen.'

Just being indoors was treat enough – and a flush was very heaven.

'When at last, in 1966, my parents were allocated a bungalow with an inside toilet,' reported James Payne, 'my father would just go and sit in there because, as he said, he'd reached the ripe old age of sixty, and this was the first time ever in his life that he'd had the pleasure of sitting on an inside toilet.'

At the end of the twentieth century we're practically all flushers now, and wouldn't accept anything else. Daniel Bostel should, on this reckoning, be accounted one of our secular saints – and those who would like henceforth to worship at his shrine should know that his business continued to thrive after his death; that Bostels built the Winter Garden on the Palace Pier before the First World War and converted the Dome and Pavilion into an Indian hospital during the conflict itself; and that the firm is still in business locally today.

But was there no alternative to all this water?

[1 2]

Long Live the Privy!

'We wouldn't have invented water closets
if we'd known what we know today'

Lem Putt would be proud of him. There it stands: the perfectly proportioned bog, with its comfortable wooden seat and its through ventilation. True, the door opens outwards and the exterior is single-coloured (Lem favoured a two-toned effect so that you could see it better at night), but Bill Cutting is without doubt the champion privy builder of Sullington County.

And there the levity ends, because what we have under test here in Sussex is a privy for the future. Its uses in this country may be limited, but its benefits for the developing world could be phenomenal.

'Our one great civilising invention has been the WC,' says Bill – and then he makes the startling remark quoted above.

'The average household flushes a third of its water down the toilet,' he explains, a veritable torrent of figures reminding us not only that he's an engineer but that he was once the managing director of Southern Water Services. 'We've spent 150 years treating sewage and now, thanks to the European Waste Water Directive, we'll have to spend billions of pounds as a nation getting the system up to standard. That's crazy.'

A shortage of water is our problem, and Bill's privy – a prototype built recently in response to the Romanian crisis – uses no water at all. 'The idea was to develop something that's cheap both to produce and to run.'

The construction could hardly be more simple. The shell can be put together on site using the cheapest material available. Bill's contribution, delivered as a flat-pack, consists of a plywood

Bill Cutting devised his Bio Bog for the Worthing-based charity Link Romania – but it may one day be seen all over the developing world.

89

seat above a plastic box. There's a chimney vent, the air at present being pulled through by a 12-volt electric fan which experiments may prove to be an unnecessary extra. A trap door at the side gives access to the box.

The privy is designed for family use, with an average of five 'serious' visits a day. You do your business, drop in a handful of sawdust and let the bacteria naturally present get to work on it. The technology has already been used on a grander scale – by the Caravan Club among others – but this is the first attempt to apply it to the average household.

'Of course, people in Britain aren't going to give up their lovely WC for a privy, however well it works. And this wouldn't do for, say, a beer festival, where you get crowds of men pouring out gallons of liquid over a short space of time.'

Bill has found, though, that the prototype sent to Romania has been a great success 'and the people who're using it say they're delighted.'

Forget those visions of grandad carrying an evil-smelling bucket to the vegetable patch. This privy produces a fine-crumbed, odourless soil which can be applied as a low-grade fertiliser without any health risks at all.

But isn't it reminiscent of the Reverend Moule's earth closet, now a museum piece?

'The sawdust is used as a bulking agent, and it keeps the smell down, too, as the earth did. But the difference is what happens next. The heat generated by bacterial action evaporates the urine and breaks down the material in the box, so that it's pretty rapidly reduced. The same process was taking place in Moule's apparatus, but earth itself wasn't as good a medium. It didn't break down fast enough, so you needed to empty the container much more frequently and therefore before the bugs had finished their work.

'Earth privies are a real source of disease in the developing world, particularly in villages which have no mains water. It doesn't take much imagination to realise that going to the toilet

All you need add is a little sawdust . . .

. . . and then you take the contents out through this little trap door – and share it with your friends!

in a hole in the ground at the bottom of the garden and getting your water from a shallow hole at the top is a recipe for disaster.

'In fact, there are occasional outbreaks of hepatitis A and other gastro-intestinal diseases, and the finger of suspicion points directly at the very unsatisfactory toilet system.'

Managing Bill's privy is a doddle. Since your deposits will inevitably create a pyramid of waste, you rake over the contents occasionally to flatten them out and spread them evenly over the bottom of the box. If they appear to be drying out, you pour a little water in; if they're too wet, you scatter more sawdust. Every nine months or so you'll need to clean the privy out, but this is simply a matter of opening the flap to reach the box.

'And you leave a third of it in there, so that the bacteria can continue their good work on what comes later. A third of it can be used as compost.'

If your maths are sound, you'll realise that you still have a third left. Any guesses what you do with it? Yes, *you give it to some- one else.*

'It's rather like the ginger beer plant,' laughs Bill. 'It's a living thing, so it keeps on working. If you have one of these privies in a Third World village, you give your surplus to someone else who's just installing one. Believe me, he'll be very pleased to have it.'

And why not, since there's no smell – and that means that these little loos don't have to sit down the end of the garden path as in days of yore. They can be tucked into spare spaces right next to the house, or even inside it.

'That's what I mean about the direction we took being a mis- take,' Bill says. 'Of course it's great to flush everything away and out of sight, but we're not really getting rid of it at all. We're sending it to the water company to deal with, and that's where the problems start.'

It may have virtually disappeared from the Sussex landscape, but there's life in the old bog yet. Long live the privy!

A Privy By Any Other Name

> 'I remember that it was called the bog or closet
> by the men and the privy by the women.'

Mary Shelton's intriguing comment, offering an interesting line
of enquiry for any future privy explorer, was the sole reference
during my researches to naming by gender.

Certainly the privy has had a great many names over the centuries – some understandably coarse and others desperately
prim – and I am pleased to append a roll of honour which you
will simply have to imagine as being soft and perforated.

A 'certain' place
Aster room
Biffy Bog
Boghouse
Bombay
Chamber of commerce
Chamberlain pianos ('bucket
    lav')
Chuggie
Closet
Cludgie
Comfort station
Crapper box
Crapphouse
Crapping castle
Dike
Dinkum-dunnies
Doneks
Dubby

Dubs
Duffs
Dunnekin
Dunnick
Dyke
Garderobe
Go and have a Jimmy Riddle
Go and have a Tom Tit
Going to pick the daisies
Going to see a man about a
    dog
Going to stack the tools
Going to the George
Going to the groves
Gone where the wind is always
    blowing
Gong
Gong house
Heads

Here is are
Holy of holies
Home of rest
Honk
House of commons
House of office
Houses of parliament
Jakes
Jericho
Jerry-come-tumble
Karzi
Klondike
Knickies
Larties
Latrine
Lav
Lavatory
Little house
Loo
My aunts
Nessy
Netty
Out the back
Petty
Place of easement
Place of repose
Place of retirement
Reading room
Round-the-back
Shit-hole
Shittush
Shooting gallery
Shunkie
Slash house
The backhouse

The boggy at the bottom
The bush
The dispensary
The dunny
The grot
The halting station Hoojy-boo
  (attributed to Dame Edith
  Evans)
The house where the emperor
  goes on foot
The hum
The jakers
The jampot
The japping
The John
The lats
The long drop
The opportunity
The ping-pong house
The proverbial
The Sammy
The shants
The shot-tower
The sociable
The tandem (a two-holer)
The thinking house
The throne room
The watteries
The wee house
The whajucallit
Thunder box
Tivvy
Widdlehouse
Windsor Castle
'Yer Tiz'

ACKNOWLEDGEMENTS

My indebtedness to the scores of kind folk who plied me with memories, anecdotes, sketch maps, press cuttings, privy plans, photographs and postcards will be obvious enough to any reader of this book. Many are mentioned in the text, and I trust that they will do me the further kindness of thereby regarding themselves as thanked in the most material way possible.

Many more have helped 'behind the scenes', and (with apologies to any I've missed) I would like to record my gratitude to Graham, Amy, Barbara, Clive and David Butler, Stuart Cannings; Jeanne and Ken Cunningham, Pat Eaves, Norman Edwards, Edwina Alison Fox, Vida Herbison, Brian Johnson, George Pixley, Eddy Powell, Colin Richards, Clive Robbins; Ian Smith, Miss M. F. Ullrich and Simon Wright.

Most of the photographs in the book are my own, but I am profoundly grateful to several people who supplied illustrative material: David Robinson for the postcard of the Hurstpierpoint postbox privy (page 21); Jane Seabrook for copying the old photograph of East Hoathly church (page 82); and Southern Water for the photograph of the Brighton sewers (page 84).

Several authors and publishers have kindly allowed me to quote freely from their books: Dirk Bogarde and Random House UK Ltd for the extract from *A Postillion Struck by Lightning*; the ever-generous Bob Copper, whose memories, and those of his father Jim, are sprinkled through that wonderful Sussex trilogy *Early to Rise*, *A Song for Every Season* and *Songs and Southern Breezes*; and Charles Sale and Putnam & Co for extracts from *The Specialist*.

My thanks, too, to the *Evening Argus* and Harry Cawkell for permission to reprint the extract from 'Nature's Way' (page 22), and to News International for permission to reproduce the Lewes privy accident report from *The Times* (page 60).